W9-DAU-877

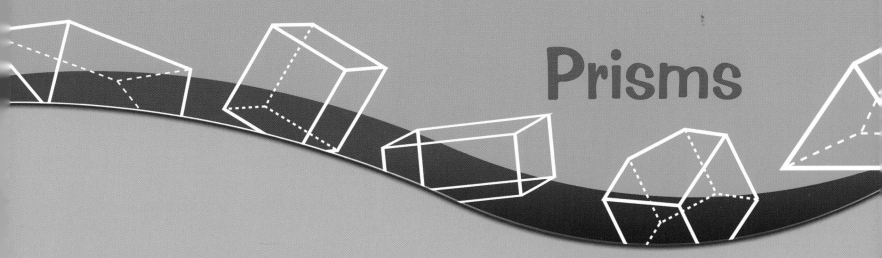

Prisms

Bonnie Coulter Leech

The Rosen Publishing Group's
PowerKids Press™
New York

To Bill and Billy—the wind beneath my wings

Published in 2007 by The Rosen Publishing Group, Inc.
29 East 21st Street, New York, NY 10010

Copyright © 2007 by The Rosen Publishing Group, Inc.

All rights reserved. No part of this book may be reproduced in any form without permission in writing from the publisher, except by a reviewer.

First Edition

Editor: Kara Murray
Book Design: Elana Davidian
Layout Design: Greg Tucker

Photo Credits: Cover © Michael Newman/PhotoEdit; p. 5 © Alan Schein/zefa/Corbis; p. 6 © Clayton J. Price/Corbis;
p. 12 © Randy Faris/Corbis; p. 13 © K. Hackenberg/zefa/Corbis; p. 14 © Michael Dunne, Elizabeth Whiting & Associates/Corbis;
p. 16 © Roger Ressmeyer/Corbis; p. 21 © age fotostock/Superstock.

Library of Congress Cataloging-in-Publication Data

Leech, Bonnie Coulter.
  Prisms / Bonnie Coulter Leech. — 1st ed.
     p. cm. — (Exploring shapes)
  Includes bibliographical references and index.
  ISBN 1-4042-3498-5 (lib. bdg.)
  1. Prisms—Juvenile literature. 2. Geometry, Plane—Juvenile literature. 3. Geometry, Solid—Juvenile literature. I. Title. II. Series.
  QA491.L44 2007
  516'.156—dc22
                            2005034255

Manufactured in the United States of America

Contents

Solid Figures

Look around your classroom at all the different shapes and sizes. These are likely to be books, desks, cabinets, and chalkboards in your classroom. They all have different shapes. They are all different sizes. Look around your city or town. You will see many objects of different shapes and sizes.

All these different objects are solid figures. Solid figures are not flat. They do not lie in a plane. A plane is a flat surface like a sheet of paper. A plane has length and width, just like a sheet of paper, but the length and width of a plane continue forever.

Two-dimensional figures lie in a plane. The figures have length and width. A solid figure is **three-dimensional**. Figures that are three-dimensional have not only length and width, but also they have height.

This picture shows Times Square in New York City. From the buildings to the billboards to the streetlights, all the objects you see in this picture are solid figures.

Prisms

In **geometry** three-dimensional figures are called solids. One example of a solid that we study in geometry is the prism. You may also have heard the word "prism" in science class. In science a prism is a piece of glass that breaks up light into the seven colors of the rainbow. When light passes through a prism, you can see red, orange, yellow, green, blue, indigo, and violet.

In geometry a prism is a three-dimensional solid. The top and bottom of a prism can be any kind of shape, but they will always have the same shape and the same size as

Scientific Prism

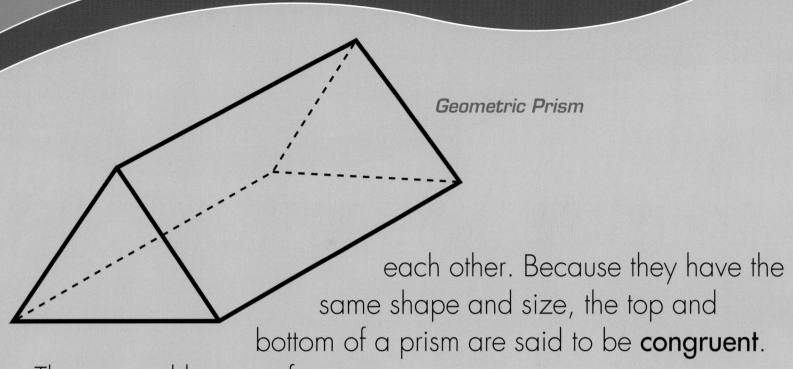

Geometric Prism

each other. Because they have the same shape and size, the top and bottom of a prism are said to be **congruent**.

The top and bottom of a prism also never meet or cross each other. They are **parallel**.

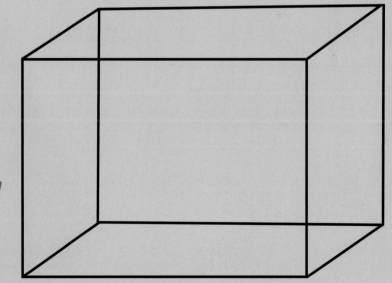

Geometric Prism

7

The top, bottom, and sides of a prism are all called **faces**. The faces of a prism are **polygons**. Polygons are two-dimensional, closed figures with three or more sides. Polygons are closed because all their sides meet another side.

The faces that make up the sides of a prism are called **lateral faces**. The lateral faces of a prism will always be a polygon with four sides. In a **right prism**, the lateral faces will be rectangles. A rectangle has four sides. In this book when we say "prism," we will be talking about a right prism. The lateral faces will always be rectangles.

The faces that make up the top and the bottom of a prism are called bases. The bases of a prism can be any kind of polygon. A prism does not always sit or lie on its base.

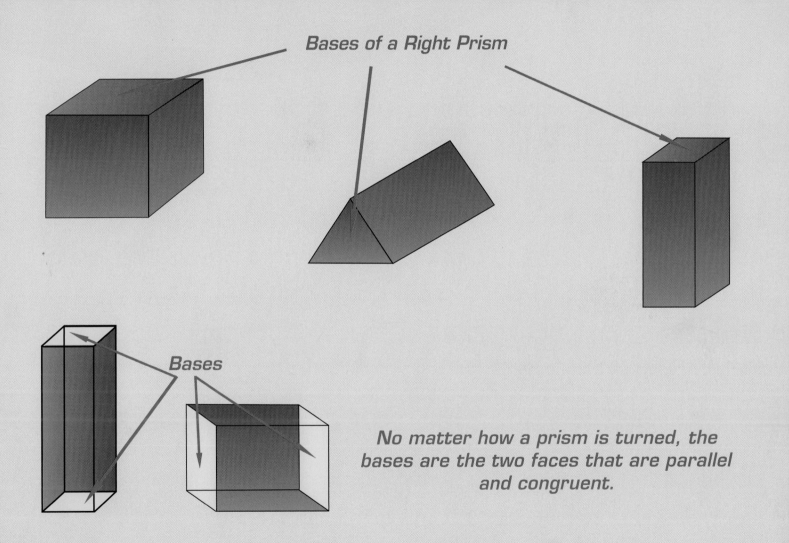

Bases of a Right Prism

Bases

No matter how a prism is turned, the bases are the two faces that are parallel and congruent.

Prisms are named according to the shape of their bases. If the polygon that makes up the base of a prism is a triangle, then the prism is called a triangular prism. Because a triangle has three sides, the prism would also have three sides, or lateral faces. Like the lateral faces of all right prisms, the lateral faces of a triangular prism would be rectangles. A triangular prism has a total of five faces. It has two triangular bases and three rectangular lateral faces.

If the polygon that makes up the base of a prism is a rectangle, then the prism is called a rectangular prism. A rectangular prism has a total of six faces. It has two bases and four lateral faces. All the faces of a rectangular prism are rectangles.

A polygon that has five sides is called a pentagon. If a prism has two bases that are pentagons, then the prism is called a pentagonal prism.

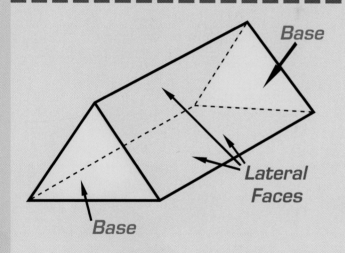

Base

Lateral Faces

Base

A right triangular prism has bases that are triangles. The three lateral faces, or sides, are rectangles. The bases are shown in yellow.

A pentagonal prism has bases that are pentagons. The bases are shown here in yellow.

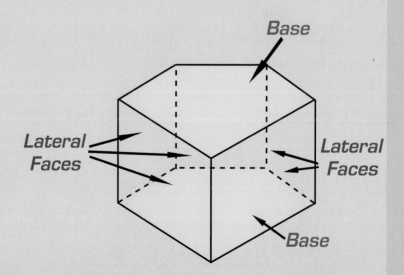

Base

Lateral Faces

Lateral Faces

Base

Cubes

Prisms can have many different types of polygons for their bases. If a prism has a square base, then the prism is a square prism. A square is a rectangle with all four sides congruent. A square prism with four square lateral faces is a **cube**.

All six faces of a cube are squares. All six faces of a cube are congruent to each other. Have you ever played with a pair of dice? Dice are cubes. All their faces are congruent.

Where the faces of the cube meet are the edges of the cube. Each edge forms one side of two congruent square

Pair of Dice

faces. Therefore, all the edges and all the faces of a cube are congruent. A cube is a special rectangular prism because all its sides are congruent.

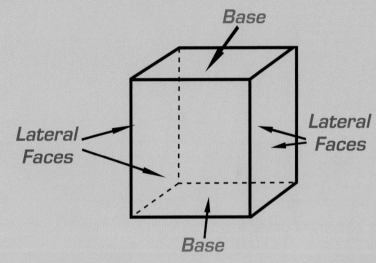

A Rubik's Cube, shown here, is made up of 26 small cubes with differently colored faces that fit together to form one larger cube. The object is to make each face of the whole cube one color.

This is a cube. All six of the faces are congruent. Depending on how the cube is turned, any two of its faces that are parallel could be the bases of this cube.

When two planes meet, they form a line segment. A line segment is a part of a line with two endpoints. Look at the corner of a room, where two walls meet. The corner of a room is a line segment. Look at the place where the wall meets the floor. They meet in a line segment, too.

The wall and the ceiling meet to form a line segment.

Two walls meet to form a line segment.

Where two faces of a prism meet, they form a line segment. That line segment is called an edge. The place where two edges of a prism meet is called a corner. A corner is also called a **vertex**. The word for more than one vertex is "vertices."

The **lateral edges** of a prism are the edges that are formed by the lateral faces of the prism. Remember that the lateral faces of a right prism are rectangles. The lateral edges are the sides of the rectangles that make up the lateral faces. They are not the sides that help form the base.

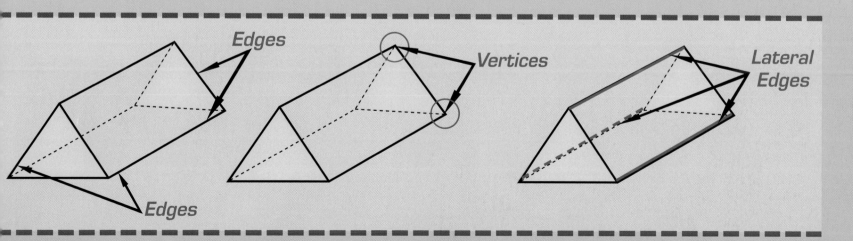

Edges *Vertices* *Lateral Edges* *Edges*

Have you ever ridden in an airplane and heard the announcement of the **altitude** at which the airplane was flying? Have you ever looked at a map and seen the altitude of a mountain written on the map? The altitude of an airplane is how high the airplane is flying. The altitude of a mountain is the height of the mountain.

Altitude 12,388 feet
(3,776 m)

Mount Fuji, shown here, in Japan has an altitude of 12,388 feet (3,776 m). A mountain's altitude is the height of the mountain. The altitude is the distance from sea level to the top, or summit, of the mountain.

Prisms and other solid figures have height. The height of a prism is called the altitude. The altitude, or height, of a prism is the distance from one base to the other base. The altitude of a right prism would also be the length of one of the sides of the lateral faces, or lateral edges. The lateral edge of a right prism is the same length as the altitude.

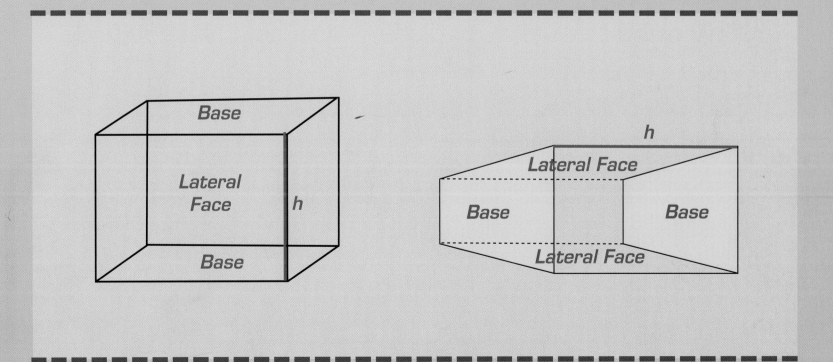

Have you ever taken a box and opened it up so that it lies flat? When you flatten a box, it becomes easy to see each part of the box. If you cut along the edges of a prism and unfold the prism into a flat shape, you have made a **net**. A net is a two-dimensional, or plane, figure that can be folded to make a three-dimensional figure.

Nets are patterns, or two-dimensional pictures, of prisms. Many differently shaped nets can be a pattern for one prism. The shape of the net depends on how the prism is unfolded. For example, there are several different ways that you can cut the edges of a cereal box so that you get a flat box. Therefore, there are several different nets that can be used to form a three-dimensional cereal box.

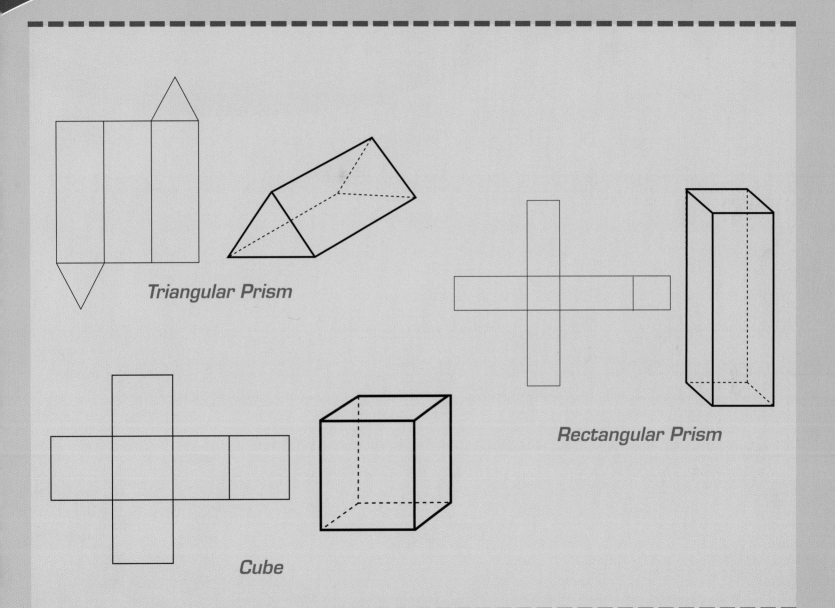

Triangular Prism

Rectangular Prism

Cube

Surface Area

The **surface area** of a prism is the sum of the areas of all the faces of the prism. If you take a box that holds a birthday present and wrap the box, you have covered the surface area of the box. If you painted the walls, the floor, and the ceiling of a room, then you would have painted the surface area of the room.

A net is helpful in finding the surface area of a prism. If a prism is flattened into a net, then you can easily see each of the polygons that make up the faces. You could then count the number of squares that go into each of the faces. When you add up the number of squares in every face, you will get the surface area of the prism.

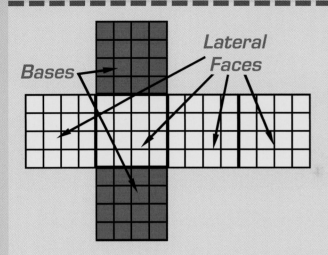

Bases

Lateral Faces

Let's find the surface area of this cube. Count the squares in each of the faces on the net shown. In a cube all six faces are congruent, so the bases and the lateral faces all have the same measurement. The faces of the cube shown here all measure 4 squares by 4 squares. They all have an area of 16 squares. To find the surface area of this cube, we must add 16 + 16 + 16 + 16 + 16 + 16, or 16 x 6. It equals 96. This cube has a surface area of 96 squares.

This girl is painting the walls of a room. If she also paints the floor and the ceiling of the room, she will have covered the surface area of the room with paint.

Prisms Around Us

There are prisms all around us, and they can have many uses. Honeybees use hexagonal prisms to make their honeycombs. In Sweden and other places where it snows a lot, triangular prisms are used to make A-frame houses. An A-frame house is shaped like the letter A. People build houses in that shape because snow cannot build up on the roof of a triangular house. The snow will only slide off.

We can find prisms in many different places. Look on your teacher's or principal's desk, and you may see a name plate in the shape of a triangular prism. Juice boxes are perfect shapes for holding juice for lunch. They also fit very well in lunch boxes. Juice boxes and lunch boxes are both rectangular prisms. Prisms are everywhere in our three-dimensional world.

Glossary

altitude (AL-tih-tood) The height above Earth's surface.

congruent (kun-GROO-ent) Having the same measurement and shape.

cube (KYOOB) A solid figure with six congruent square sides.

faces (FAYS-ez) Any of the polygons that form the sides and bases of a solid figure.

geometry (jee-AH-meh-tree) A type of math that deals with straight lines, circles, and other shapes.

lateral edges (LA-tuh-rul EJ-ez) In a solid figure, the line segments formed where the sides of the solid meet.

lateral faces (LA-tuh-rul FAYS-ez) In a solid figure, the polygons that form the sides.

net (NET) A two-dimensional pattern that can form a three-dimensional figure when folded.

parallel (PAR-uh-lel) Being the same distance apart at all points.

polygons (PAH-lee-gonz) Two-dimensional, closed, many-sided figures.

right prism (RYT PRIH-zem) A prism whose altitude is one of its edges and whose lateral faces are rectangles.

surface area (SER-fas AYR-ee-a) In a solid figure, the sum of the areas of each of the figure's faces.

three-dimensional (three-dih-MENCH-nul) Having length, width, and height.

two-dimensional (too-deh-MENCH-nul) Able to be measured two ways, by length and by width.

vertex (VER-tex) The point where two lines or line segments meet.

Index

Web Sites

Due to the changing nature of Internet links, PowerKids Press has developed an online list of Web sites related to the subject of this book. This site is updated regularly. Please use this link to access the list:
www.powerkidslinks.com/psgs/prisms/